John Wesley Hill

A decade of curious insects:

Some of them not describ'd before, shewn in their natural size, and as

they appear enlarg'd before the Lucernal microscope, in which the solar

apparatus is artificially illuminated

John Wesley Hill

A decade of curious insects:
Some of them not describ'd before, shewn in their natural size, and as they appear enlarg'd before the Lucernal microscope, in which the solar apparatus is artificially illuminated

ISBN/EAN: 9783337714079

Printed in Europe, USA, Canada, Australia, Japan

Cover: Foto ©berggeist007 / pixelio.de

More available books at **www.hansebooks.com**

A

DECADE

OF

CURIOUS INSECTS:

SOME OF THEM NOT DESCRIB'D BEFORE:

SHEWN IN

THEIR NATURAL SIZE;

AND AS THEY APPEAR ENLARG'D BEFORE

THE LUCERNAL MICROSCOPE;

In which the SOLAR APPARATUS is artificially illuminated.

With their HISTORY, CHARACTERS, MANNERS,
and PLACES of ABODE;

On TEN QUARTO PLATES, and their Explanations.

DRAWN AND ENGRAVED FROM NATURE.

By J. HILL, M. D.
MEMBER OF THE IMPERIAL ACADEMY.

LONDON:
Printed for the AUTHOR, in St. James's-Street.
And Sold by B. WHITE, in Fleet-Street; P. ELMSLY, in the Strand;
PARKER, in Cornhill; BALDWIN, in Pater-noster-Row; RIDLEY,
St. James's Street; and J. BALFOUR, at Edinburgh.

MDCC.LXXIII.

Ladies who may chuse to paint thefe Infects themfelves may have Sets of the Cuts on Royal Paper printed pale for that purpofe.

INSECTS,

ENGRAVED FROM NATURE.

CLASS I.

THOSE WHICH HAVE
FOUR GAUZY WINGS,
AND
A WEAPON IN THE TAIL.

BY Gauzy Wings, we underſtand ſuch as are thin, tender, and tranſparent: not cruſty, as the Beetles; nor leathery, as the Crickets; nor duſty, as the Moths and Butterflies; but *clear*. Such are the wings of common Flies.

GENUS

G E N U S I.

S A W - F L Y.

T E N T H R E D O.

Character of the Genus.

The Mouth is form'd of Jaws; and has no Trunk.

The Scutcheon, has two small, distant, elevated points, on its hinder part.

The Wings lie plain; but are a little puff'd up, and uneven.

The Weapon at the tail is short; and form'd of two plates, jagged like a Saw; and hollow'd lengthwise in the Female. Plain in the Male. Plate 1. *a b c d.*

All two-wing'd Flies have a pair of Plummets behind their Wings; rising from under a bloated Scale. Those swellings in the Saw-Fly seem to be such Scales not open'd; and never disclosing any Plummets.

Nature does all things regularly; and makes her advances by equal and gradual degrees: and this seems her gradation from the two-wing'd to the four-wing'd Classes of Insects; the first in which the Plummets cease.

We shall find throughout her universal regions, that creatures differ by equidistant steps from one another; and that this difference, this advance of Species above Species, is all her laws allow. All real knowledge of her works is, and for ever will be, confined to this; the knowing and establishing the differences of one Species from another: Classes, and Genera, tho' useful, are arbitrary; ideas of mens minds; that exist not in nature.

To know these characters of difference, is all: but the parts which mark them; the greater, as well as the lesser; are so imperfectly seen in the smaller Insects, that their names, or kinds, often cannot be known; nor does the mind perceive the wonders of the Creator display'd in these his creatures. 'Tis therefore they are here represented both in their natural size; and as they appear before a small, but distinct magnifying power: and that way only they can be either well known, or justly admir'd.

I. MOURN-

I.

MOURNING SAW-FLY.

TENTHREDO LUCTUOSA.

Plate 1.

Character of the Species.

The ANTLERS have feven joints, and are all the way of a thicknefs,

The HEAD and Trunk are red; the Body is black.

Plate 1. *a.*

This pretty, quiet, melancholy Fly is found among Alder Plantations; and is often fatally entangled in the clammy juice, that oozes from their Leaves. I caught it this laft May, by the road-fide, near Uxbridge.

Its HEAD is of the fineft fcarlet;
The *Eyes* are blue..
The *Antlers* are of a dufky brown, and hairy.
The *Feelers* fhort, and pale.
The *Mouth* is arm'd with hard and crufty jaws.

Its TRUNK is fcarlet above, and of a ruddy brown below.
The *Scutcheon* is of a deeper red,
The *Points* on it are blue..

The BODY is coal-black above, and greyifh black below.
Its *Rings* are divided by lines, form'd of a deep brown membrane joining them.
Its *Air-holes* are of a dead brown.

The LEGS are grey; they are all of a length, and have two claws.

The WINGS are of a pale yellowifh brown, with little yellow rifings on the ribs, and an edge of deeper yellow,.

The TAIL is of a deep brown,

The

The STING or Saw which terminates it, is flatted, and thin, and of a chefnut brown; faw'd in the Female, plain in the Male. On prefiing the body of the Fly between the fingers, it may be forc'd out farther, a little from the vent.

This is the *Tenthredo-Alni*, of the Syftema Naturæ of Linnæús. Perhaps alfo, it is the Tenthredo Ovata of the fame work; for Infects are not fo numerous, as 'tis the cuftom now to think them: and colour, tho' an obvious, is no certain character among thefe creatures: in fome it differs with the feafon; in others, with the fex; in all, it glows according to the creature's health and vigour: in moft, it is exalted in the time of courtfhip, as the feathers on the necks of fome Fowls; and in fome, it fades, and is loft utterly in dying, as the colours of many fifhes.

This pretty Fly rifes from a yellow Worm with a black head, and twenty little feet; frequent in fummer on the Alders, and bury'd under ground all winter for its change: in May we fee it perfect.

'Twere well if we knew all the Infects, as this is known; but 'tis only a fmall part that have been trac'd fo thoroughly; where they have, it makes a great addition to their hiftory: but where the eye has not diftinctly feen it, 'tis beft to be filent. They who relate their errors and conjectures, under the feeming face of knowledge, deceive, and are deceiv'd.

H. MOTTLED

Tenthredo

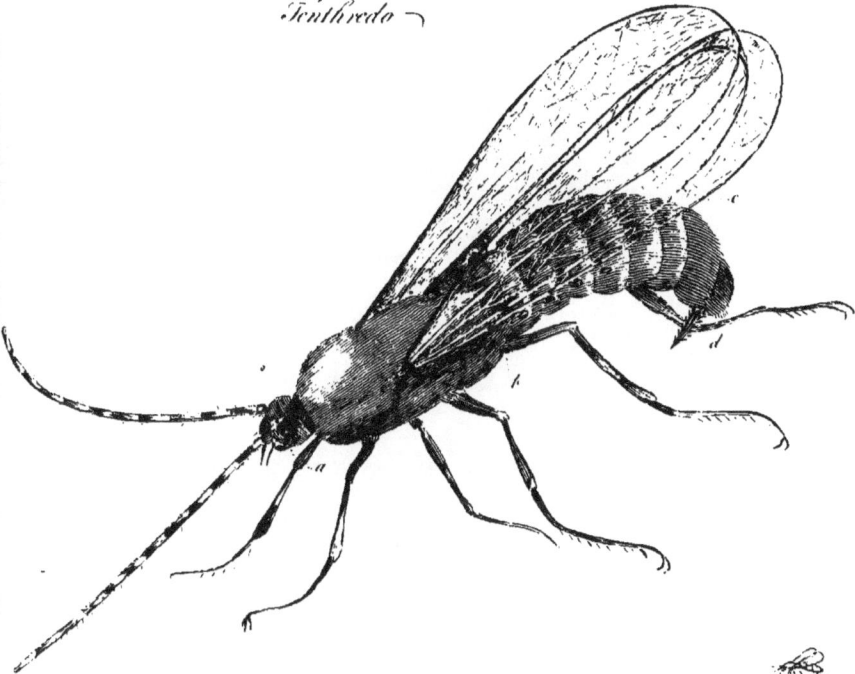

Mourning Saw-fly

Tenthredo luctuosa

II.

MOTTLED SAW-FLY.

TENTHREDO VARIEGATA.

Plate 2.

The ANTLERS have more than twenty joints; and grow fmall to the point.

The HEAD is blue; the Trunk is deep grey, mottled with yellow; the BODY is black.

Plate 2

This is a very ftrange and delicate Fly: 'tis found in damp woods and moors in Auguft and September.

The HEAD is of a fhining blue.
The *Eyes* are green.
The *Antlers* are amber-colour'd.
The *Feelers* fhort, and brown.
And the *Jaws* of a yellow brown.

The TRUNK is of an iron-grey, mottled with irregular fpots of gold, like the womens tambour-work in embroidery.
The *Scutcheon* is entirely raven-grey.
The *Points* on it are black.

The BODY is coal-black above, and raven-grey below.
The *Lines* dividing the rings are brownifh.
The *Air-holes* are black.

The LEGS are of a fine bright yellow, with black claws.

The WINGS are brown, with a dufky edge.

The TAIL is amber-colour'd.

I received

I received this pretty creature by an accident from Scotland : the Duke of Athol found, this Autumn, in an oak-wood near Dunkeld, a Whortleberry-Shrub with white fruit : a thing not known before in Britain. His Grace did me the honour to fend me fome growing Plants of this fmall Shrub, for the garden of her Royal Highnefs the Princefs Dowager of Wales at Kew ; and upon one of them came feveral of thefe Flies wrap'd up alive.

It feems the Tenthredo Sylvatica of the Syftema Naturæ of Linnæus.

The Fly is Female, that is here defcrib'd ; the Male has no Saw, for he has no ufe for it. Nature has given that inftrument to the Female, to cut a way into a growing Vegetable ; and there to lodge the eggs : which pafs through the hollow made by the two fides or plates of the Saw.

The Male has in the fame place, a kind of Forceps, or Pincers, which he can thruft out, and with them feize the Female. Thus in larger animals, where the Female has teats for fuck, the Male has a refemblance of them for conformity.

When the Female lays her eggs, there goes with them an acid mucilaginous juice, which perverts the courfe of the Sap in the Plant, and makes it grow into a kind of gall : this operates as foon as it is iffu'd, and cicatrizes the part the Saw had wounded.

One may fee bubbles of this juice always left upon the Plant ; it is foft and clammy : the wound is oblong, and crooked, and the part becomes black as if burnt : the egg increafes in bignefs to twice or more than that, after it is lodg'd in the Plant ; nor is this ftrange fince it has no hard covering.

Tenthredo

2

Pl. 2

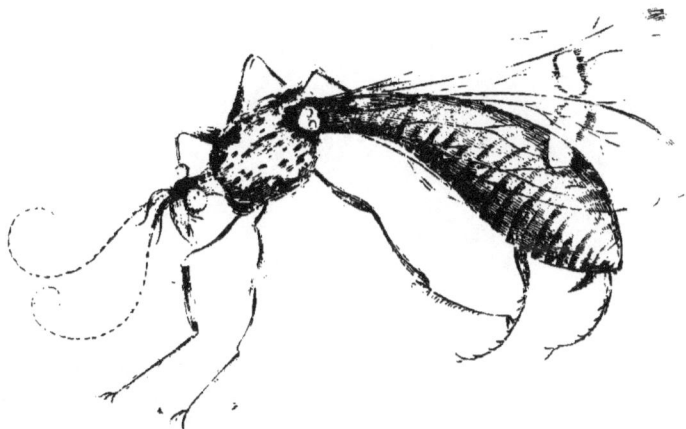

Mottled Saw-fly

Tenthredo Variegata

G E N U S   II.

T H E   S A V A G E.

S P H E X.

Plate 3.

Character of the Genus.

The MOUTH is form'd of oblong Jaws, without Trunk, or Tongue.

The WINGS lie smooth, and perfectly even.

The ANTLERS have ten joints.

The WEAPON at the tail is simple, sharp, and hollow ; and does not appear, except the Body be press'd; or the creature strikes with it.

Nothing can be so provident as this creature for its young; nor any thing so savage, as the means it uses for that purpose. The manner of living is different in the various Species ; and so is the general form of the Body : the present and succeeding figure will shew this difference of shape : yet all the while the Genus is, and must be allowed the same, because the Characters in all the distinctive parts agree. And in the same manner, tho' the place of shelter, and the course of life, be utterly different; yet the same manners appear innate, and inherent in both.

They agree in being the fiercest of all Flys ; they will attack Insects much larger than themselves ; and this, whether they be defencelels, or arm'd, as they are, with a Sting. The Strength in all this savage kind is great ; their Jaws are hard, and sharp ; and in their Sting is a poison, suddenly fatal to the creatures with whom they engage. The Savage seizes hardily on the creature he attacks : and gives a stroke of an amazing force ; and then falls off, as if himself were kill'd : but 'tis to rest from his fatigue, and to enjoy his victory. He keeps a steady eye on the creature he has struck, 'till it dies, which is in a few minutes ; and then drags it to the nest, for the young. The number of other Insects these destroy, is scarce to be conceiv'd ; the mouth of their cave is like a Giant's of old in romance ; strew'd with the remains of prey : the Eyes, the Filament that serves as Brain, and a small part of the contents of the Body, are all the Savage eats : and he will kill fifty for a meal.

B   1. C O M B-

1. COMBFOOTED SAVAGE.
SPHEX PECTINIPES.

Plate 3.

Character of the Species.

The ANTLERS are form'd of oval Joints, and turn like rams-horns.

The FORE-FEET are form'd like combs, with three claws, and ſtiff hairs above.

The BODY is join'd cloſe to the Trunk.   Plate 3. *a b.*

This ſtrong and fierce, tho' heavy Fly, lives in caverns of the earth, in the ſides of hills and cliffs; and in holes made in the mud-walls of our little villages. I received this from the North of Ireland, where the mud-walls of one of the cabins on the ſide of a hill, was wrought into the appearance of a Honeycomb, by the multitudes of theſe creatures.

Its HEAD is of a cheſnut brown.

The *Eyes* are blue.

The *Antlers* are brown; but the tip of each joint is ruddy.

The *Jaws* are amber-colour'd.

The *Feelers* pale brown.

The TRUNK is black, and rough.

The *Scutcheon* is grey.

The BODY is ſmooth, and ſhining; of a ruſty iron colour, with bands of an orange yellow.

The *Air-holes* are brown.

The LEGS are of a blue grey; and the long hairs upon the fore ones, toward the feet, are yellowiſh.

The WINGS are of a pale brown.

The STING, when the creature pleaſes to ſhew it, is of a fine poliſh'd brown.

This ſeems unqueſtionably the Sphex Pectinipes of the Syſtema Naturæ.

It is drawn here, as ſhewn by the fourth glaſs of the lucernal microſcope; not magnify'd in any vaſt degree; but ſufficiently to ſhew all its parts. Creatures much ſmaller require often larger figures to expreſs their organs diſtinctly. All magnitude is comparative; and to be uſeful, the inſtrument ſhould be employ'd with juſt ſo much power as is needful for diſtinctneſs.

2. T H E

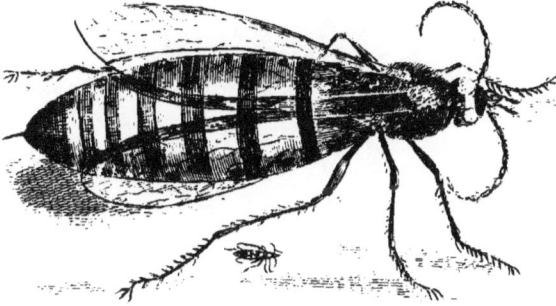

Sphex

...nted Savage   *Sphex Pectinpies*

2. THE TURNER SAVAGE.

SPHEX SPIRIFEX.

Plate 4.

Character of the Species.

The Body is join'd to the trunk by a long fmall thread.

The Antlers have ten joints; and they fpread out, and grow fmall to the point.

The Feet are jointed, and equally hairy; and have each two tces.

This ftrange disjointed creature, as it feems, lives, by choice, among men, whom it never offends; but it is beyond meafure terrible to the leffer infects: and by the fabricature of its dwelling, it might become an object of furprife and wonder; tho' there were nothing more to recommend it to our notice.

I received this particular Fly from Peterborough in Northampton-fhire, where it had form'd its cells in the mud-wall of a fmall cottage, juft under the edge of the thatch; dry, warm, and fhelter'd from the weather.

The preceding kind lives in a mere cave of its own making; a fimple, oval hole, with a fmall opening, and larger within: this had turn'd its dwelling in a clofe fpiral form, and polifh'd the infide fo well, that it had the afpect of one of the fpiral fhells we fee in cabinets, when faw'd open: about the mouth of this, was form'd a kind of funnel, covered with legs and wings of flaughter'd Infects; and juft within this mouth ufually fat the inhabitant watching what came by; for the ftrange ftructure of his body made him lefs fond of flying far; left half of it fhould be left behind him.

Befide the aftonifhing havock of this creature among the Infects, on which it preys, there is a part of its hiftory ftrangely replete with horror: it has been obferved, that while the Savages are fo deftructive of other creatures, they have a wonderful attention to their young; and this, by a courfe of Providence unknown to us, any more than by the term inftinct, appears in all their actions, even before thofe young are born.

In

In the preceding kind, the eggs are laid in the back part of the cavern where the creature lives; evenly arrang'd; and when the time of their hatching is near, the Fly brings in a number of flaughter'd Infects, for the food of the expected young ones: fhe then clofes up the mouth of the hole with mud, and her care is over. When the young worms hatch, they find their food ready; and when they have eaten their fill, they reft, and take their change into the Fly.

But this creature lays her eggs in the body of a living Caterpillar: they hatch, and eat that creature up, even while itfelf is feeding: at their appointed time they hatch: and 'twas long a wonder among the curious, how a Caterpillar produced this Fly, inftead of a Butterfly, or Moth; or how one Infect chang'd to many.

The HEAD of this creature is of a chefnut brown, with a fhade of blue.
The *Eyes* are black, and large.
The *Antlers* are of a ruddy brown.
The *Feelers* are blackifh.
The *Jaws* are hard, ferrated, and yellow.

The TRUNK is of a ruddy brown.
The *Scutcheon* is yellow.
The *Thread* which faftens the two parts together, is alfo yellow.

The BODY is of the colour of rufty iron; but there is a fkin of yellow covering part of it from the end of the thread.
The *Air-holes* are black.

The LEGS are partly brown, and partly yellow.

The WINGS are of a dufky brown.

The STING is yellow.

The drawing of this, as of the former, is not greatly magnify'd; the fame fourth glafs was us'd to it; the creature being naturally of a fize nearly big enough to fhew its own particularities; and always here the lefs magnifying is wanted, the lefs is us'd.

GENUS

Sphex
2

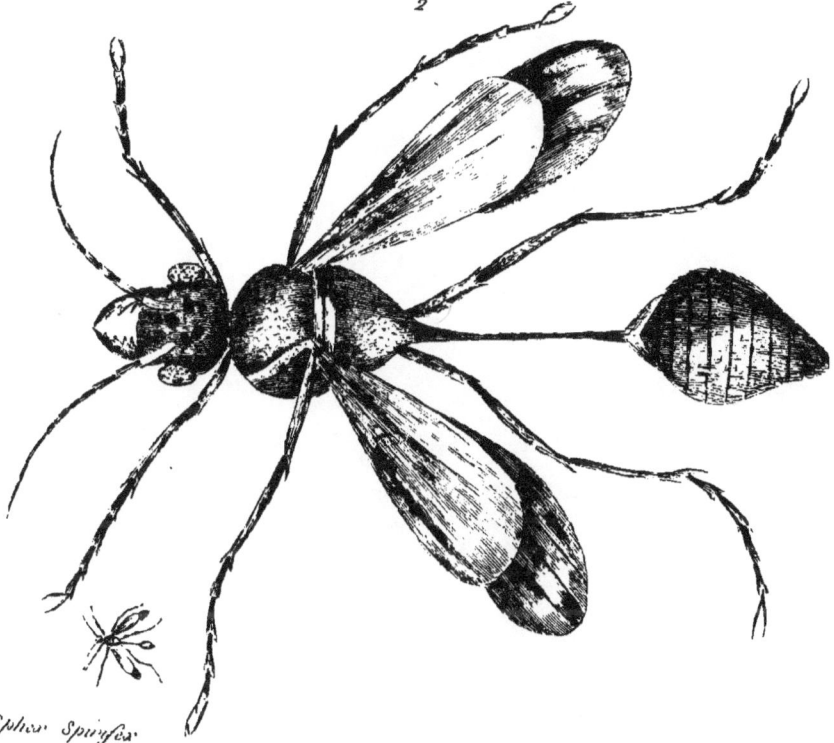

Sphex Spirifex

The Turner Savage

G E N U S III.
A N T - E A T E R.
M Y R M E L E O N.

Character of the Genus.

The MOUTH is form'd of Jaws, with two long Tufks.
The ANTLERS are club fafhion'd.; and there are four long Feelers.
The WINGS hang down.
The TAIL is arm'd with a pair of Knippers, in the Male.

1. T H E   G R E Y   A N T - E A T E R.
M Y R M E L E O   F O R M I C A R U M.

Plate 5.

In many of the wing'd Infects, their prior form of the Worm, or
Reptile, rifing immediately from the egg, demands a fhare of our attention, with the Infect in its more perfect and more beautiful appearance; in the prefent kind, our greateft admiration is demanded
in that lefs perfect ftate. The Butterflies arife from Caterpillars;
the Beetles from fix-footed Worms; and the Dragon-flies from Infects
without Wings, which fwim about in water. The creature under
confideration here, approaches to the Dragon Fly in kind; and in
its figure, in the Reptile State; being a broad and bloated hexapode;
but inhabiting the dryeft earth.

It is known that birds and beafts of prey can endure great and long-
continued hunger; the fierceft moft. This creature, ally'd to the
Savages in its manners, can alfo bear their abftinence: La Hire, of the
Paris Academy, obferv'd about fourfcore years ago, that the creature
could bear a feven months faft; 'twas to him we owe the firft notice
of this Infect; fo well defcribed foon after by Vallifnieri, and Pou-
part; and fo much fpoken of, and fo poorly underftood, by the petty
retailers of natural knowledge fince.

The Reptile State of this pretty Fly, known by the name Formicaleo, is a coarfe Infect, of a pale yellow, ftreak'd with brown, and
varied with fome black tufts of hairs; but ufually it is fo covered with
dirt, that it looks brown: its habitation is under ground; it forms a pit,
like a funnel, of dry duft, and lies conceal'd in the centre of it, to
catch the Ants, or other little creatures, that fall into it.

Its Head is broad, and flat, and has a pair of Tufks, or Horns; or,
by whatever name we may call parts unknown to larger animals; thefe
are

are fharp, open, and hollow : with thefe he pierces the bodies of Infects ; with thefe he draws in their juices for his food ; and when
that is done, they have an elaftic force, by which they throw the
carcafe far away. He retires under ground to feed, and juft rifes to
throw the refufe out of his pit ; then repairs its injuries, and waits
for the next chance. Thefe pits are about three inches wide ; the
creature leaves them, and makes new ones at his pleafure :' and in
this ftate he always lives many months, fometimes two years, before
he turns into the Fly, now to be defcrib'd.

MYRMELEO FORMICARUM.
THE ANT-EATER FLY.
Plate 5.

The ANTLERS are compos'd of twenty-four joints, and grow larger
to the tip.

This is a large Fly, not fwift in its motions, but fierce and deftructive ; even in a degree equal to that of the Reptile, from which it
fprings : it plays about the bufhes in the meadows of France and Italy, in the latter part of fummer, and will feize on almoft any thing
it can catch.

Its HEAD is of a chefnut brown.

The *Eyes* are vaft, and green.

The *Antlers* are of a deep brown.

The *Feelers* are long, and dufky ; there are four of them.

The *Jaws* are yellow, hard, and fharp ; and the two Tufks are brown.

Its TRUNK is of a greyifh brown, with a gilded variegation.

The *Scutcheon* is blueifh.

Its BODY is of a pearly grey, deep, and not elegant.

The *Lines* or *Rings* are black.

The *Air-holes* are edg'd round with brown.

Its LEGS are fhort, ftrong, and ruddy, with long dark hairs.

Its WINGS are grey ; and in the particular Fly before me, there are
four fpots of a dufky brown upon each of the upper ones, and two on
each of the under: I fay, in this particular Fly, which is from Italy;
for there are more, or fewer, or none, in thofe from other places.

The Knippers at the Tail are horny, and chefnut colour'd.

This is the Fly of the famous Formicaleo, the Myrmelea Formicarum of the lateft writers.

GENUS

Myrmela.

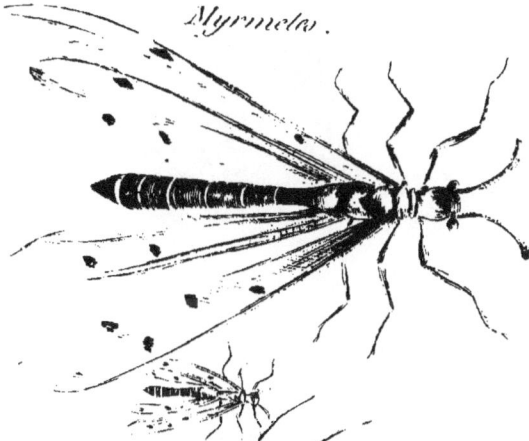

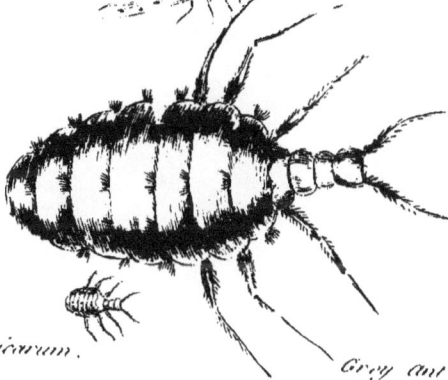

Myrmela Formicarum.

Grey Ant Eater.

G E N U S IV.

G A L L F L Y.

C Y N I P S.

Plate 6.

Character of the Genus.

The Mouth is form'd of oblong Jaws, and has no Trunk.

The Weapon at the Tail is fpiral: it is naturally hid; and only Females have it.

They have an ink in Norway, rude, and unciviliz'd, as the country in a great meafure is, which excels that of all the world in colour, clearnefs, and permanency. Some letters, I had the honour to receive from the Bifhop of Bergen, gave me an opportunity of obferving this; and an enquiry into its compofition, produced an Hiftory of the fucceeding Fly; frequent in France and Germany, as well as there; and I think not unknown in England: but 'tis with caution we muft afcertain the Species of thefe leffer animals. Colour has been taken in as an effential character, but it changes here.

Our oaks give food and lodging to a multitude of Infects: I think there are not lefs than forty-feven creatures of this rank, perfectly diftinct in fpecies, and of many Genera, that live in, and on it.

We fee upon the oak-leaves in our woods in June, round balls, as big as nutmegs, green, with a blufh of red, and foft to the touch: thefe are the leaf galls with which the Norway ink is made; and we have nearly the fame Fly that makes them. They arife from a wound made by that Infect, who lays an egg there; and in their centre there is a fmall cavity, within which the Worm lives, that, after a time, hatches into this Fly.

The common galls, with which the common ink is made, and which are alfo of fo great ufe in dying, rife from the young fhoots of the oak, not from its leaves; and they are hard and woody. Thefe are more numerous; and as they fall with the leaves in Autumn, they might be collected eafily in great quantity; and may perhaps be of value, by improving more than one article in Commerce.

1. THE

1. THE OAK LEAF GALL FLY.

CYNIPS QUERCUS FOLII.

The ANTLERS are hoop'd, and have a thick extremity.

The TRUNK is bloated, and ſtreak'd.

This Inſect I received from Norway, where the oaks were in a manner covered with it, in the months of July, and part of Auguſt, 1768; and in our own oak-woods I have ſeen ſuch a Fly frequently ; and found it, tho' differing a little in colour, perfectly the ſame in all its characters. It is a ſtrong, coarſe-made, and not very handſome Fly; and is ſlow and heavy in its motions : and is generally found ſitting on the under part of a leaf, with its wings ſpread out flat.

Its HEAD is of a ruſſet brown.

The *Eyes* are blue.

The *Antlers* are mottled, of black and cheſnut colour.

The *Feelers* are brown.

The *Jaws* are ſharp, jagg'd, and brown.

Its TRUNK is of a raven grey, very beautifully ſtreaked with white.

Its BODY is coal black.

The *Rings* dividing the Joints are brown.

The *Air-holes* are duſky.

Its LEGS are grey, with ſtiff black hairs, and black toes; and the thigh is black.

Its WINGS are of a pale brown, with a tinge of olive-colour; the Veins on them are deep grey.

Its TAIL is perfectly black.

The *Sting* is cheſnut brown.

With us the Trunk and Body of this Fly are quite black ; otherwiſe there is no difference between ſome I caught this year in Buſhy Park ; and thoſe I had from Norway.

Cynips.

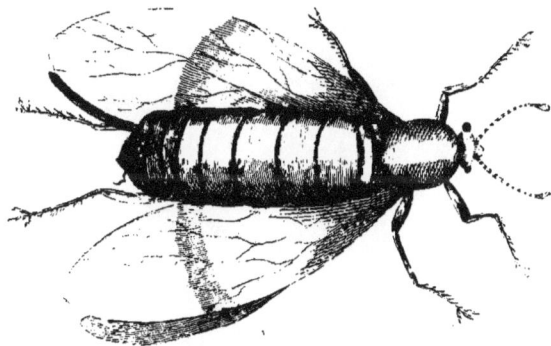

Cynips Quercus folii.

Oak leaf Gall Fly.

GENUS V.

DAY-FLY.

PHRYGANEA.

Plate 7.

The MOUTH is oblong, and without Teeth.

There are no *Feelers*.

There are two large *Studs* upon the Head, juft above the eyes.

The WINGS are carry'd ftanding upwards, and are unequal in fize.

The TAIL has briftles annex'd to it.

The Day-Flies are an inoffenfive race; born to pafs thro' their little ftage of being, the prey to a thoufand enemies; but hurtful to no creature: they live about waters, in which they breed; and in their Fly ftate have fo fhort a term, that it has been the fubject of feparate hiftories, by Naturalifts, and Emblems for moral writers. The name, Day-Fly, arifes from their living in that ftate but one day: but in many of the fpecies, even that period is much longer than is allowed.

The particular kind firft to be figured and defcribed in this place, never burfts from its Reptile ftate, till about fix o'clock in a fummer evening; and never lives to fee the next fun rife.

Five hours complete its little fpan of life; in the which time, if it efcape the Fifh, the Dragon-Flies, and Reed Sparrows, (for all are after it) it copulates with the Male; depofits its impregnated eggs in the waters; and dies before the cold of midnight.

But 'tis not that thefe hours are all it lives; 'tis in thefe only it enjoys the air: but the Worm hatched from the egg of this Fly lives, and feeds heartily in the waters, enjoying a much longer date, and that in more fecurity; for it covers its tender frame with a motley cafe of its own conftructing; and gormandizes unfufpected, and unfeen; for one, or fometimes nearly for two years.

C   1. THE

3. THE WHITE-WING DAY-FLY.

EPHEMERA CULICIFORMIS.

Plate 7.

The ANTLERS have a multitude of knotted joints, and grow fmaller to the point.

The BRISTLES of the Tail are hard, and firm.

This is a fwift-wing'd Fly, abundant about running waters, in the months of June and July; where it becomes the food of a multitude of fifhes : many leap at it as it drops toward the water; and others watch the reeds and rufhes near the fhore, and take it with more eafe as it is dropping its eggs. This was caught in July laft, near Efher.

Its HEAD is of a dufky brown.
The *Eyes* are green.
The *Studs* are jet black, and fhine.
The *Antlers* are of a chefnut brown.
The *Mouth* is a kind of amber-colour'd beak.

Its TRUNK is of a tawny brown, with a brighter fpot in the middle.
The *Scutcheon* is nearly white.

Its BODY is of a dead brown.
The *Rings* are pale.

Its LEGS are of a greyifh, or afh colour.

The WINGS, tho' not decorated as the Moths, or Butterflies; yet have a peculiar and wonderful prettinefs : they are of a pearly white, mottled here and there, and clouded as it were with the fame colour, only thicker, or lefs pure.

The TAIL is pale brown.
The *Briftles* are ruddy.

2. ROCK

Ephemera.

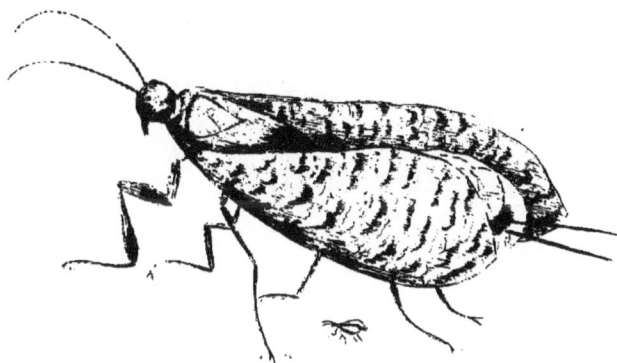

Ephemera Culiciformis.

White Wing'd
Day Fly.

2. ROCK DAY-FLY.

EPHEMERA RUPESTRIS.

Plate 8.

The ANTLERS have a multitude of clofe-connected joints, and grow all the way fmaller to a point.

The BRISTLES at the Tail are fhort, and weak.

Nothing can be ftranger than the hiftory of this Fly, which came to my knowledge by an accident laft year; and, I believe, has not been obferved by any writer.

On a ftone obelifk, erected before a houfe in London, to fupport the lamp, I obferved feveral oblong, greyifh tubes, or cafes, running in various directions; fome ftrait, and others a little bent. I fhould have fuppofed them the tubuli, or cafes of Sea Worms, petrify'd, as is frequent in many kinds of ftone; but that thefe obelifks carried very plainly the marks of the chiffel; and the little tubules I obferved were wrought over them; and therefore evidently had been formed after the ftone was work'd.

The fingularity of this, caufed me to direct a fervant to pick off fome of them; which he attempted in vain: he found them as hard as the reft of the ftone, and fixed to it with great firmnefs: with the help of a hammer, fome few were at length got off; and I found nothing fhelly in them; but that they were mere ftony tubes, form'd of the matter of the obelifk, in fmall granules, cemented clofe.

In breaking feveral others, I at length found in fome, the creature which had form'd them for its houfe and fhelter: this was a little yellowifh Worm, with a black head, and a number of fmall, fhort feet. It ufually refided in the bottom of the tubule; but came out at pleafure.

On fome cobwebs, about the upper part of the obelifk, I found, among other Infects, the remains of two or three Flies, feeming either of the Day-Fly, or Moth kind; but too imperfect for me to afcertain the Species.

C 2   Thus

Thus rested the matter for that time; but my curiosity being roused by the strangeness of the incident, I examined large masses of stone, wherever I saw them, this last year; and happening to be in Buckinghamshire in July, I found the whole mystery explained.

Several large stones that stood in water at their bottom, tho' dry enough above, were covered with grey, stony tubules of this kind; and about one of these masses, on the evening of the 18th of July, I found more than fifty, of the Fly, to the Worm of which they owed their origin. 'Tis a very pretty creature, and in all respects of the Day-Fly kind.

Its HEAD is hoary, and of a strong fine green, with a black round spot on its centre, shining at the summit.
The *Eyes* are black as jet.
The *Studs* are brown.
The *Antlers* are scarlet, long, thrust strait forward, and usually cross'd.
The *Mouth* is dusky.

Its TRUNK is of a lovely green, and is join'd to the body by a kind of neck, which is also of a velvety green.
The *Back* is beautifully variegated with streaks and dots of gold.
The *Scutcheon* is of a lighter green.

Its BODY is thick, and green, and is connected to the Trunk almost without a division.

Its LEGS are of a pale brown.

Its WINGS of the same pearly grey with the preceding; but elegantly vein'd, and clouded with a pale blue, and a light brown.

The BRISTLES are amber-colour.

I believe this Fly lives no longer than one evening in its wing'd state: the Females, among those I saw, were very busily depositing eggs in all the cracks and crevices of the stones. The Worms hatched from these, make the stone tubules for themselves, and probably live in them one or two years.

INSECTS.

Pl 8

Ephemera

2

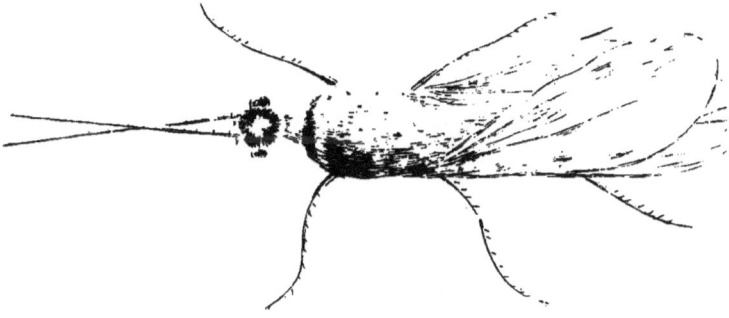

Ephemera Rupestris

Rock Day Fly

I N S E C T S.

C L A S S II.

Thofe which have four feathery wings.

BY feathery wings, we underftand fuch as are form'd, in appearance, as of the feathers of birds : each wing confifting of only one fuch feather; tho' fometimes fplit, or divided.

G E N U S I.
C H I N C H.
A L L U C I T A.

Plate 9.

The ANTLERS are compos'd of a few oval joints; and the extreme one runs out into a point.

The TAIL is fplit, and hairy.

The FEATHERS, which are plac'd as wings, confift of jointed ribs, and thin flat plates fet regularly on them.

The Chinches are a race of Infects fo extremely fmall, that they have in a great meafure efcaped obfervation. Few of thofe who have ftudied thefe fmall objects, have feen any of them ; and from fuch as have, very little of their nature is to be learn'd, for they have only cafually come before the eye : and thofe microfcopes, by the affiftance of which they have been examined, and figured here, have not been known till lately.

The creatures in the Infect world, to which the Chinches approach the neareft, are the feather-wing Moths ; but from thefe they differ abundantly in the ftructure of their Antlers, and the fhape of their body, their motions, and peculiar formation of their Tails. Thofe Moths have been called Phalenæ Alucitæ ; and the latter term therefore alone feems the moft familiar and intelligible name for thefe.

The feathers which compofe, or rather which are the wings of the Chinches, tho' they very much refemble the plumes of birds in appearance, are in reality very different, and have nothing truly feathery in them. They are compos'd of a hollow jointed rib, not unlike fome of the Corallines ; and the hairs, or plumes, as they feem, which rife from them, are flat, thin, conic fcales.

1. THE

1. THE STRAW-COLOUR'D CHINCH.

ALLUCITA PALLIDA.

Each wing is compofed of one diftinct undivided feather.

Plate 9.

This is a creature very ftrange in its nature, and hiftory; and which once came as ftrangely before me. A ftudious gentleman, very fubject to the head-ach, which he, and his phyfician, both attributed to great attention; fneezing one day with violence, as he was writing, faw fome atoms a moment afterwards upon a fheet of white paper that lay upon his table; and they plainly moved : he doubled up the paper, and brought it to me : when we laid a parcel of thefe moving particles before the lucernal microfcope, they appeared of the fize and figure reprefented at Plate 9; and were in continual motion; vibrating their Antlers, fhaking their Wings, and turning up their Tail to their Heads, in the manner of Earwigs, but with an incredible fwiftnefs.

'Twas palpable they had been difcharged from his nofe; and 'tis very eafy to fee whence they were thrown, and to underftand how they might have caufed intolerable pain, while they were thus raifing and moving their irritating hairs, and feathers, upon a part where the very fubftance of the brain is almoft naked.

I had feen the fame Species inhabiting the Flowers of the Plant Mignonette; and on afking, found he had that Plant in his chamber.

The HEAD of this creature is lemon-colour'd.
Its *Eyes* are of a delicate blue.
The *Studs* over them deep black.
Its *Antlers* are of the paleft brown, but ruddy at the bafe of each Joint.
The *Feelers* are pale, and fmall.

Its TRUNK is of a pale ftraw-colour.
The *Scutcheon* has a tint of greenifh.

Its BODY is very pale ftraw-colour.
The *Rings* dividing it are whitifh.

Its LEGS are pale brown, but deeper at the joints.

Its WINGS are whitifh, with a dufk of brown.

Its TAIL is amber-colour'd.

Alucita

Pl. 9

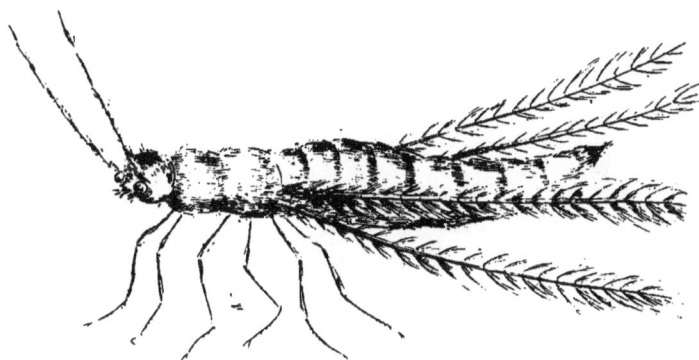

Alucita Pallida

The Straw-colour'd
Chinch

2. THE TAWNY CHINCH.

ALLUCITA FULVA.

thrip corny nbi lio no n . Ret'i l .

Plate 10.

Each wing is compofed of two feathers, rifing from a fimple bafe.

This little creature lives in the hollows of the Flowers of Plants, as the preceding; and feems calculated to do at leaft as much mifchief, being fully as fmall; more covered with Hairs, or fhort Briftles; and to all appearance, both as to its ftructure, and motions, able to drive them into the tender membranes with more force: a bigger bodied, and more robuft creature.

It is a wanderer, and lives in a manner at large in gardens: 'tis fondeft of the fweeteft Flowers: the Damafk-Rofe is often full of them, and the ftock July-Flower, and Wall-Flower; nor does it difdain the Lupine, and the Pea.

The characters, and ftructure of the Chinches, are in no Species feen more diftinctly than in this.

Its HEAD is of a dull yellowifh brown.
The *Eyes* are large, and of a fiery red.
The *Antlers* are firm, elegantly jointed, a little hairy, and very fharp at the points; and they are of a pale brown.
The *Feelers* are fhort, and dufky.

Its TRUNK is brown, covered with pale hairs, and variegated with a dead yellow.
The *Corcelet* is paler.

Its BODY is of a tawny brown, divided by dufky rings, and covered thick with fhort, firm, whitifh hairs, which it can raife at pleafure.

Its LEGS are of a dufky brown, ftrong jointed, and have forked, hard, horny toes, and fome very ftiff hairs upon them.

It

Its Wings are of a pale dusky yellow; the rib of them is seen beautifully jointed; and is palpably hollow.

The Scales, which make what are call'd their hairs, or feathery substance, are very numerous, narrow, and sharp, both at the edges, and the point; and they make a great appearance, because, each being compos'd of two feathers, the creature seems, in comparison with the preceding, to have four wings upon each side.

The creature has, when living, all the motions of the preceding kind, but stronger; as its Body seems more muscular : and the power it has of raising its Bristles, as the Porcupine, must add greatly to the irritation, when it is got into a tender part.

Whether, and how far, head-achs are to be traced from these little creatures, is a subject yet for more enquiry; but 'tis worth the pains. Many have this pain from the smell of Flowers. Some have been found dead, with quantities of violets, and other Flowers, in their chamber. Physicians have attributed these deaths to the powerful odour of those Flowers; but that they should be owing to these creatures, is much more probable.

Whether they do, or do not, ever fly far from the Plants they inhabit, I have not yet found; but in the stillness of the night, it would not be strange if they should : and that they hover round them I have seen : for placing a strong light, and a great convex glass, near a pot with a growing Lupine, in a dark chamber, I have discovered the air, all about the tops of the Plant, in a manner filled with them, moving like motes in a sun-beam.

Chinch.
2

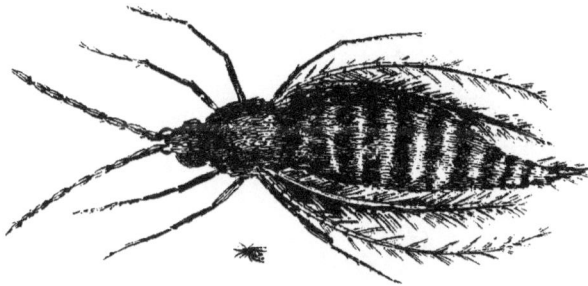

Alucita fulva.   The Tawny Chinch

I N D E X.

I N D E X.

F I N I S.

www.ingramcontent.com/pod-product-compliance
Lightning Source LLC
Chambersburg PA
CBHW022026190326
41519CB00010B/1619